George Michael Edebohls

Total Extirpation of the Uterus

Cases Illustrating Various Indications for and Different Methods of

Performing the Operation

George Michael Edebohls

Total Extirpation of the Uterus
Cases Illustrating Various Indications for and Different Methods of Performing the Operation

ISBN/EAN: 9783337811198

Printed in Europe, USA, Canada, Australia, Japan

Cover: Foto ©berggeist007 / pixelio.de

More available books at **www.hansebooks.com**

TOTAL EXTIRPATION OF THE UTERUS.

CASES AND REMARKS.

[From the Transactions of the New York Obstetrical Society, March 1st, 1892.]

DR. G. M. EDEBOHLS presented the specimens from a case of
CASE I. *Puerperal Pyosalpinx and Intra-peritoneal Abscess.
Chronic Pelvi-peritonitis. Abdominal Pan-
hysterectomy. Recovery.*

Mrs. I. S., aged twenty, was married at fifteen and had given
birth to three children. She had been mildly ill since her second
confinement, nearly four years ago. Her symptoms since then
had been pelvic pains, vesical and rectal disturbances and fever.

On admission to hospital, February 25, 1892, a large exudate
was found filling the entire pelvis, distending Douglas' sac, and
reaching upwards to near the umbilicus. Neither tubes, ovaries,
or even uterus could be made out in the mass. A fluctuating
point was discovered to the right and another to the left of the
median line. From the former a yellowish, very albuminous
serous fluid, and from the latter a few drops of shreddy pus, were
obtained on exploratory puncture. During the five days of observation previous to operation, the evening temperature reached
103½°, the morning temperature being nearly normal. The
other well-known symptoms of sepsis existed.

Dr. Edebohls had performed abdominal section that afternoon.
After opening the abdomen and separating the adherent omentum
and intestines from the pelvic viscera, a mass presented containing uterus, both tubes and ovaries, all covered with a dense
fibrous exudate one to two centimeters in thickness, the contained organs being indistinguishable from each other even when
brought into plain view with the aid of the Trendelenburg posture.
An abscess, containing about three hundred grammes of pus,
occupied Douglas' sac, and the adjacent parts of the peritoneal

cavity. After emptying the abscess it became evident that the mass anterior to it contained the uterus and adnexa, nothing else lying between bladder and rectum. It being impossible to distinguish where uterus ended and the adnexa began, an exact salpingo-oophorectomy was clearly impossible, and the mass was removed by ligaturing with catgut down along either pelvic wall until the vagina was reached, and cutting out the entire uterus with the adnexa between the ligatures. After flushing and drying of the peritoneum, two rubber drainage tubes were carried from the abdominal wound through the pelvis out into the vagina, and the pelvic cavity packed with iodoform-gauze, the end of which was also led into the vagina.

Dr. Edebohls presented the specimen as parallel with that presented by Dr. Polk at the last meeting of the Society. In both instances salpingo-oophorectomy for pyosalpinx was contemplated. In both cases the limits between uterus and appendages were difficult, if not impossible, to recognize. Dr. Polk attempted the separation of the appendages but found himself obliged to extirpate the entire uterus, removing it with the appendages, in order to control hemorrhage. Dr. Edebohls saw the futility of attempting to separate the appendages and proceeded at once to extirpate the entire diseased mass comprising the uterus, tubes and ovaries.

March 14th. Gauze removed on the third and drainage tubes on the seventh day. With the exception of an intercurrent attack of acute catarrhal pneumonia, convalescence was uneventful. (Patient left hospital, a well woman, six weeks after operation.)

[From the Transactions of the New York Obstetrical Society, March 21, 1893.]

DR. G. M. EDEBOHLS presented three uteri removed by operation, the indication and the method in each case being different.

CASE II. *Fibrosarcoma of Vagina and Left Broad Ligament. Perineotomy and Cœliotomy. Removal of Tumor, Uterus and Appendages, and entire Left Broad Ligament. Recovery.*

R. S., aged thirty-six, married, the mother of six children, the youngest being five years of age, was admitted to St. Francis' Hospital, October 31, 1892. For four years past has suffered much from backache and pains in lower abdomen, constipation, cardiac palpitation, dyspnœa and general weakness. Her periods are regular, the flow appearing every four weeks and lasting two to three days.

She presents a wide diastasis of the recti abdominis allowing the escape of the intestines beneath the skin and superficial fat. The tubes and ovaries are normal in size, non-sensitive on pressure and prolapsed backward with the retroverted fundus uteri. Uterus is normal in size, retroverted in second degree and can be readily replaced by bimanual manipulation. Cervix lacerated bilaterally, considerably hypertrophied, with an unhealthy look of the granulations and nodules thickly covering the lips of the lacerations. One of these nodules was excised and sent to Dr. J. W. Brannan, pathologist of the hospital, who, after examination, reported it as simple hyperplasia of the cervix.

On November 8, 1892, curettage of the uterus, amputation of the cervix and shortening of the round ligaments were performed at one sitting.

On December 15, patient was ready to leave the hospital, when on examination previous to discharge a suspicious nodule, *not found at the last vaginal examination, ten days previously*, was discovered. This nodule measured a little over a centimeter in diameter and was situated high up in the left half of the recto-vaginal septum, at the base of the broad ligament. It involved the posterior vaginal wall, through which it had just began to ulcerate. The rectum was still freely movable over the posterior aspect of the nodule. From the upper end of the nodule a cord-like thickening could be felt extending upward and outward into the left broad ligament to beyond a point where it could be traced by the fingers.

The clinical features were so evidently those of a malignant neoplasm that extirpation of the growth was immediately determined upon. On the assumption that the growth was probably secondary to malignant changes in the uterus, a supposition sustained by the appearance of the cervix before amputation, it was resolved to remove the uterus also.

The operation was performed on the following day, December 16, 1892, and was commenced as a perineotomy. A liberal elliptical incision, with its concavity forward, was carried across the perineum from one tuber ischii to the other between the anal and vaginal orifices. This incision was deepened and the rectum separated from the vagina until the nodule was reached. The nodule was first carefully freed from the rectum, and then cut away with a border of healthy vaginal wall, two centimeters wide all around, attached.

The induration extending from the upper border of the nodule.

was next followed outward and upward in the broad ligament
and removed with a good bit of the peritoneum as far as sight
and touch would go, the work being all the time in dangerous
proximity to the left ureter.

A section of the removed nodule was made at this juncture and
the malignant character of the neoplasm established beyond
doubt.

It was next attempted to retrovert the uterus and to bring it
down through the large opening in the peritoneum by hooking
two fingers over the fundus from behind. The recently shortened
round ligaments, however, held the uterus so well forward as to
render this manœuvre impossible of execution. Dr. Edebohls felt
confident that but for the shortened round ligaments it would
have been a very easy matter to remove the uterus through the
peritoneal and the perineal wounds.

The insertion of the round ligaments into the cornua of the
uteri might, although with great difficulty, have been cut from
below and the uterus thus liberated. The necessity of further
removal than could be accomplished from below, and the infiltration in the left broad ligament indicated cœliotomy. The
abdomen was therefore opened by a ten centimeter median incision above the pubes and, in the Tendelenburg position, the entire
uterus with both tubes and ovaries were easily removed in one
piece. The left broad ligament was now carefully dissected out
entire, clean out to the pelvic wall, leaving none of its contents
but the ureter, which on inspection and palpation seemed perfectly normal.

It was found impossible to cover the large gap in the bottom of
the pelvis with peritoneum. The extensive raw surfaces were
therefore packed with iodoform gauze, the ends of the strips being
led into the vagina, and the abdominal incision was completely
closed.

The patient was again placed in the dorsal position, the gauze
packing of the pelvis adjusted from below so as to drain through
the vagina, and the perineal wound accurately closed by numer-
ous interrupted sutures of silkworm gut. The separated vagina
and rectum were simply placed in apposition, and what was left
of the vaginal tube was allowed to take care of itself. The operation lasted nearly two hours, the time being fairly evenly divided
between the cœliotomy and the perineotomy.

Very great shock, lasting for two days, and a mild attack of
acute catarrhal pneumonia, beginning on the third and ending on

the eighth day, followed the operation. With these exceptions patient made a good and rapid recovery and was out of bed at the end of two weeks.

The gauze packing was renewed four times during the first week and then discontinued. Both the cœliotomy and the perineotomy wounds healed by primary union.

Patient left hospital on January 5, 1893, twenty days after operation, very anæmic in appearance, with the wound in vaginal vault well closed, but an inflammatory induration about six centimeters in diameter still occupying the region of the removed left broad ligament.

As presented to the Society this evening, over three months after operation, she is the picture of health, and no induration of any kind can be palpated in the pelvis. The perineotomy scar is scarcely recognizable. Since leaving hospital patient has been doing all the heavy work of a large family.

The removed specimens were carefully examined by Drs. Brannan and Freeborn, and the tumor pronounced a fibro-sarcoma. Sections of the uterus in all directions failed to reveal any evidences of malignancy. Unfortunately the cervix, removed by amputation five weeks previously, and which was most open to suspicion, had been thrown away.

Nearly all that has been published concerning the operation of perineotomy saw the light of day in 1889. Before and since that year the literature of the subject has been so meagre as to amount to practically nothing.

Perineotomy was proposed by Otto Zuckerkandl (Wiener med. Presse, 1889, No. 7,) for extirpation of the rectum, exposure of the prostrate and total extirpation of the uterus. The advantages he claimed for the operation were control, by the eye, of the parametria, the posterior vaginal wall, the rectum, and the relation of the uterers to malignant, chiefly carcinomatous, infiltrations.

At the third annual meeting of the German Gynæcological Society, in 1889, Wiedow related some cases of pelvic abscess approached and opened by perineotomy. Frommel reported upon the method, citing several cases of total extirpation of the uterus by perineotomy. He preceded the operation, differing from Zuckerkandl, by first circumcising the cervix and separating the bladder in the manner usual in vaginal hysterectomy. Saenger related a case of dermoid cyst of the cavum subperitoneale pelvis removed by perineotomy, and alluded to a previous case of Mickulicz-Trzebicky in which a cyst was also removed by way of the perineum. In the discussion Hegar claimed priority both as to proposal and execution of total extirpation of the uterus by way of the ischio-rectal cavity. He claimed for the procedure the advantage of being able to see what you are doing.

The incisions made by different operators varied. Zuckerkandl recommended a curved transverse incision with the concavity backward. Frommel adopted this incision. Saenger's incision was sagittal instead of transverse, extending from the right labium majus directly backward to two centimeters behind the anus.

Dr. Edebohls had employed the curved transverse incision with the concavity forwards. The posterior vaginal wall, the rectum, the parametria with the ureter, and Douglas' sac were all exposed to the eye in a most satisfactory manner, never to be obtained in an ordinary vaginal hysterectomy. He was favorably impressed with the operation on this his first trial and in a suitable case would employ it again.

He had related the case somewhat at length, because, as far as his knowledge went, it was the first reported case of perineotomy in this country.

(January, 1894. Patient remains perfectly well and free from relapse, thirteen months after operation.)

DR. EDEBOHLS also reported a case of

CASE III. *Carcinoma of Uterus, Vagina and Sacral Glands. Sacral Hysterectomy. Death on the Fifteenth Day.*

P. R., forty-three, widow. Mother of four children, the last born in 1876; no miscarriages. Menstruation began at seventeen and has been regular, the patient flowing three to four days every four weeks, until about three months ago. Since then atypical metrorhagia, not severe, offensive vaginal discharges, pains in back and abdomen and loss of flesh.

Admitted February 16, 1893. Patient pronouncedly cachectic. Tubes and ovaries cannot be distinctly palpated, owing to condition of cervix and parametria, but do not seem to be enlarged. Corpus uteri about normal in size and position, but very hard. Cervix very much enlarged and hollowed out into form of a large crater with mouth downward. Walls of the crater composed of carcinomatous tissue running up on all sides to vaginal junction and posteriorly involving the vaginal wall itself. Rectum unaffected and freely movable over involved part of vagina. Slight thickening of parametria adjacent to vagina on left side. No thickening on right. Two sacral glands, enlarged to an average diameter of nearly two centimeters can be plainly felt behind rectum. Mobility of uterus very limited, scarcely two centimeters in downward direction.

The immobility of the uterus, the infiltration of the left parametrium, but especially *the secondary involvement of the sacral glands* put vaginal or cœlio-vaginal hysterectomy out of the question. Nothing but a sacral hysterectomy could meet the indications. This was offered the patient with the proper explanations, accepted, and performed on the day following admission.

With the patient etherized and in the Sims' position, a slightly curved incision about twenty centimeters long, with its convexity to the right was made from the posterior inferior spine of the ilium over the spines of the sacral vertebræ and coccyx to within two centimeters of posterior anal margin. Removal of coccyx and a transverse strip, one centimeter wide, of lower end of sacrum. Division of prevertebral fascia. Removal of two large carcinomatous sacral glands. Separation of rectum from posterior vaginal wall and retraction of rectum to left. A small hole torn into rectum between anus and peritoneal reflection was immediately closed by running catgut suture. Tedious and difficult search for Douglas' sac, consuming some twenty minutes. Opening of peritoneum and enlargement of opening to right. Uterine body normal in size and not adherent. Small cyst of the right ovary ruptured in the attempt to deliver. Both tubes inflamed, thickened, with occluded ostium abdominale, and adherent, with the ovaries, to adjacent parts and organs. These adhesions rendered retroversion and delivery of the uterus and appendages very difficult. Arduous ligation with catgut of both broad ligaments, beginning at infundibulo-pelvic ligament and ending at vaginal vault. Difficult separation of uterus from bladder. Peritoneum now closed by running suture of catgut, the work thus far having been entirely in clean, healthy tissues. Amputation of vagina two centimeters below neoplasm and removal of uterus and amputated part of vagina. Some infiltration of bladder wall by neoplasm. Upper end of vagina left open. Wound closed except for about eight centimeters near middle. Wound cavity packed with iodoform-gauze. Two rubber drainage tubes drawn through wound and vagina, emerging behind and at vulva. Time required two hours and thirty-five minutes, the longest Dr. Edebohls had ever spent over any operation whatsoever. Patient pulseless at finish. Slow reaction.

On the third day after operation a catarrhal pneumonia, involving the middle portions of right lung posteriorly, was made out ; the temperature due thereto running as high as 103½° and persisting for six days. For the last six days of life no elevation of temperature.

The pulse between operation and death varied between 100 and 140, and was at no time of any encouraging strength.

Dressings changed on day following operation, when a localized gangrene of skin over sacrum was noticed. This gangrene spread until it involved an area seven to eight centimeters in diameter on either side of the incision and extended into incision and wound cavity.

The cavity was twice daily douched with sublimate solution, 1-2000, and tamponed with gauze.

On the eighth day the sutures were removed. The wound gaped wide, its lips being gangrenous. This same dry gangrene also covered the entire surface of the wound cavity with a withered deep black pellicle, which began to be cast off only on the tenth day.

Patient died of exhaustion and mild septicæmia on the fifteenth day after operation. No peritonitis.

Dr. Edebohls confessed that he had no special liking for either sacral hysterectomy or sacral proctectomy. They constitute, especially the former, the most difficult, formidable and trying operations in the whole range of gynæcological surgery. If it were not for the fact that sacral hysterectomy offers at least a hope of cure in some cases of uterine cancer advanced beyond all possible radical extirpation by either vaginal or cœlio-vaginal hysterectomy, he would feel disposed to condemn it altogether. As long, however, as the removal of all neoplastic formation, so surely fatal as cancer, is anatomically possible, it is the duty of the surgeon to attempt it, if the patient so elect after a fair statement of the case to her, incongenial though the operation be to the surgeon.

In the case reported the carcinomatous sacral glands could be removed in no other way than by the sacral method. This particular indication for sacral hysterectomy has not been sufficiently insisted upon by writers on the subject.

The diseased condition and the universal adhesions of the appendages rendered the operation in this instance so very difficult.

As regards the technique of sacral hysterectomy we have a large variety of preliminary incisions from which to select: Kraske's, Hegar's, Hochenegg's, the parasacral incisions, left and right respectively of Emil Zuckerkandl and Wœlfler, Rydygier's, Herzfeld's, etc., etc. In this instance Dr. Edebohls followed Herzfeld's method.

The main object, always to be borne in mind, is to do all the intraperitoneal work and to close the peritoneum securely by suture before beginning to deal with the neoplasm itself. The dirty work in infected tissues can thus all be done extraperitoneally, and the risk of septic peritonitis is minimized.

DR. EDEBOHLS also presented a patient from whom he had removed the uterus for prolapsus.

CASE IV. *Complete Prolapsus of Uterus and Vagina. Total Inversion of Cervix. Vaginal Hysterectomy, Lateral Colporrhaphy and Perineorrhaphy at One Sitting. Cure.*

M. C., aged sixty-one, married, complains of nothing save a complete prolapsus uteri which occurred suddenly while lifting a heavy basket of clothes four months previously.

On examination the entire uterus and vagina are found outside of body. A complete inversion of the cervix was associated with the prolapsus, the external os being retracted high upward upon the body so as to roll out and expose the entire interior of cervix, the well-known arbor vitæ appearance of which was beautifully demonstrated, being exaggerated by intense congestion of the mucosa.

On January 27th, 1893, Dr. Edebohls performed vaginal hysterectomy, lateral colporrhaphy and perineorrhaphy in such a manner as to remove *in one piece* both tubes and ovaries, the uterus, both sides of the vagina and the posterior half of the vulva, leaving nothing of the entire genital tract except a strip of vagina anteriorly and posteriorly, and the anterior half of the vulva. The peritoneal cavity was closed by catgut sutures and the raw surfaces of the vaginal tract and the perineum were so closed by sutures as to leave on completion of the operation a vagina 8 centimeters deep by 2.5 centimeters in diameter.

This was the first time that Dr. Edebohls had removed the uterus for prolapsus, preferring as a rule the combination of the necessary plastic operations with ventrofixation of the uterus, all to be done at one sitting. The indications for total extirpation in this case were given by the age of the patient and the unhealthy condition of the uterus, especially the cervix.

(Patient left for Ireland a few weeks after her discharge from hospital, and has not been heard from since.)

[From the Transactions of the New York Obstetrical Society, May 2, 1893.]

DR. G. M. EDEBOHLS presented three fibromatous uteri recently removed by cœlio-panhysterectomy.

The first was a specimen of

CASE V. *Fibromatous Uterus Resembling a Double Uterus* and was interesting mainly from a diagnostic point of view. The fibroma, as will be seen from the specimen, was almost the exact size of the corpus uteri itself and was attached to the uterus by a short, thick pedicle arising from the junction of the corpus and cervix on the right side, so that, to the bimanual touch, a double uterus with a single cervix was exactly simulated. The differential diagnosis, however, was made before operation, without the use of the sound, by tracing both tubes to the left half of the pelvic mass, which was thus proven to be the corpus uteri. The entire uterus with the tumor and both tubes and ovaries were removed in one piece on March 2d, 1893. The patient made an uneventful recovery.

The second specimens were a fibroma and a fatty heart from a case of

CASE VI. *Fibromatous Uterus Removed by Total Extirpation. Recovery, with Subsequent Sudden Death Due to Fatty Degeneration of the Heart.*

The patient was a single woman of thirty-seven, who had known of the existence of her tumor for about three years. The tumor, a soft myoma weighing five pounds, was removed by total extirpation on February 17th, 1893, the tumor, entire uterus, tubes and ovaries being taken away in one piece.

After a perfectly smooth convalescence, and while sitting up in bed enjoying her dinner on the thirteenth day after operation, the patient suddenly turned blue, fell back in bed, gasped a few times and died.

The autopsy revealed nothing pathological except in the heart. Both auricles and the right ventricle were distended and their walls attenuated in the extreme. The right ventricle had stopped in diastole, being hyper-distended with black, partly-clotted blood. The fatty degeneration of the right ventricle was extreme, the walls at many places, throughout their entire thickness, being converted into fat. Whatever remained of the heart musculature had undergone granular degeneration.

The case illustrated the well-known association of degenerative changes in the heart muscle with fibrous disease of the uterus, and emphasized the advisability of early removal of the fibromatous uterus, with the view of forestalling the fatal degenerative changes of the muscles of the heart.

(The specimen was referred for examination to the pathologist, Dr. Freeborn, who at a subsequent meeting reported it as one of exquisite and far-advanced fatty degeneration of the heart.)

The third specimen was from a case of

CASE VII. *Cœlio-panhysterectomy for Fibroma. Acute Intestinal Obstruction Relieved by Lavage of Stomach. Recovery.*

The specimen, a fibroma weighing four pounds, was presented mainly for the purpose of directing renewed attention to the fact that the gravest cases of intestinal obstruction following cœliotomy may be relieved, after the failure of all other measures, by washing out the stomach. The patient was a single woman of thirty and had suffered for about three years with pelvic pressure symptoms, chiefly affecting the bladder, due to the presence of a fibroma snugly fitting into and filling the pelvic inlet.

The tumor, uterus, tubes and ovaries were removed in one piece *via* the abdomen, on February 23, 1893. The peritoneal wound at the bottom of the pelvis was closed by a running Lembert suture of catgut; in doing so, the sigmoid flexure was engaged somewhat in the left end of the seam. The adbominal wound was firmly closed without irrigation and without drainage.

Incessant vomiting began immediately after recovery from the ether and continued without intermission until the fourth day, the vomited matter being of a dark-green, bilious color, and towards the end having a decidedly feculent odor. No flatus was passed, and no movement of the bowels was obtainable by any of the means usually employed in these desperate cases. Prostration was extreme, the patient appeared moribund, and preparations were being made for a cœliotomy, when Dr. Edebohls, recalling the favorable results obtained by Klotz in similar cases, from lavage of the stomach, resolved to give the method a trial.

The stomach was accordingly washed out with a weak solution of bicarbonate of soda, four to five litres of the fluid being used. The effect was magical. The vomiting ceased at once and did not again recur ; complete and permanent euphoria was established, and the case from that moment on ran a perfectly uneventful course. The bowels moved spontaneously two days later.

Klotz, of Dresden, to whom Dr. Edebohls would here express his profound obligations, in a modest article, less than a page in length, (*Centralblatt fuer Gynækologie, 1892, No. 50, Page 977.*) outlines his exceptionally large and favorable experience in the treatment of intestinal obstruction following operation. Of five hundred and sixty-nine cases of cœliotomy and vaginal hysterectomy performed during a period of ten years, thirty-one had ileus, or intestinal obstruction, following operation. Six of the thirty-one were treated by cœliotomy with two deaths; the remaining twenty-five were treated by lavage of the stomach, with three deaths.

Klotz treated all of his cases on the fourth or fifth day, washing out the stomach with four to six litres of a lukewarm salt solution. If the first lavage failed to control the symptoms, a second was given soon after, followed by the introduction, through the tube, of fifty grammes of castor oil into the stomach. In all the twenty-five successful cases thus treated vomit'ng ceased immediately, flatus was passed after two or three hours, and a stool was obtained after ten hours at the latest. He holds slight, and in the beginning easily separable, adhesions responsible for the ileus in the vast majority of cases, and ascribed the relief afforded to the increased intestinal peristalsis induced by lavage of the stomach.

With these latter views Dr. Edebohls was inclined to coincide. He considered it highly probable that in his case the upper end of the sigmoid flexture contracted slight adhesions which yielded to the peristalsis excited by washing out the stomach.

[From the Transactions of the New York Obstetrical Society, November 7, 1893.]

Notes on Seven Hysterectomies.

Dr. G. M. EDEBOHLS presented seven uteri which had been removed recently by total extirpation.

Two of the operations were performed in June of the present year, and the last five within the two weeks following the last meeting of the Society. Two uteri were removed by vaginal hysterectomy, three by abdominal hysterectomy, and two by combined abdominal and vaginal hysterectomy. He thought the specimens might be of interest because they represented about every condition and indication, excepting inversion, for which the non-puerperal uterus is removed.

CASE VIII.—*Complete Prolapsus Uteri et Vaginæ. Vaginal Hysterectomy, Bilateral Colporrhaphy, and Perineorrhaphy at one Sitting. Cure.*

The patient, a widow aged seventy-one years, was referred to him by his friend, Dr. N. G. McMaster. With the exception of her local trouble, the prolapsus, she enjoyed good health. The prolapsus was first noticed eighteen years ago, and remained partial until three years ago, since which time it has been complete.

On examination, the entire uterus and vagina are found outside of the vulva. Large excoriations upon both vagina and cervix. Perinæum greatly distended. Critical study of the case, with a view to operation, made it clear that plastic work below could alone not be depended upon to cure, but that either ventrofixation or total extirpation of the uterus must be added. The latter measure was decided upon, chiefly to avoid the necessity of opening the abdomen above the pubis in the excessively stout patient.

Operation, June 8, 1893, at the home of the patient, under ether, with the kind assistance of Drs. McMaster, Leyendecker, and Sill. The entire uterus was removed from below, together with large lateral strips of the enlarged and thick vagina, and goodly portions of the mucous membrane covering the perinæum. The peritonæum was closed by a running Lembert suture of catgut; the vaginal defects were obliterated by the buried suture of catgut in tiers (*Etagennaht*); and the perinæum was closed by sutures of silk-worm gut, applied in the manner described by Dr. Edebohls (*American Journal of Obstetrics*, October, 1890). The patient took ether badly, pulse and respiration several times failing, so that during the latter half of the operation no anæsthetic was administered. She made a good recovery, however, the enjoyment of convalescence being dampened only by the presence of a number of burns due to the overzealous application

of excessively hot water-bags before the patient rallied from the
anæsthetic.   The prolapsus remains cured to this day.

CASE   IX.—*Sarcoma of the Uterus.   Cœlio-colpo-hysterectomy.
Recovery.*

A. S., a married woman aged twenty-seven years, mother of
four children.   She had a miscarriage in July, 1892, and from
that time until she came under the care of Dr. Edebohls, eleven
months later, she suffered from constant metrorrhagia, being free
from the flow only six weeks during all that time.   She had suffered much from leucorrhœa, and latterly, when straining, had
noticed something protruding from the vulva.

She presented a cachectic appearance, and there was marked
dyrexia during three days of observation preceding operation, the
temperature reaching as high as 105.4° F.   On examination, a
globular cauliflower excrescence, some seven to eight centimetres
in diameter, was found occupying and distending the upper seg-
ment of the vagina.

The tumor originated from the cervix, and on microscopical
examination, after removal, was pronounced by Dr. J. W. Brannan
to be a spindle-celled sarcoma.   A week later, June 27, 1893, the
patient's general condition having materially improved after removal of the neoplasm, the entire uterus was removed, together
with the tubes and ovaries, in one piece, by vaginal and abdominal hysterectomy.

The operation was easily performed, the patient made an uneventful recovery, and left hospital twenty-three days after operation.   He had not been able to learn anything of her since.

CASE   X.—*Carcinoma of Cervix with Myoma of Fundus Uteri.
Vaginal Hysterectomy.   Recovery.*

B. F., a married woman aged thirty-seven years, presented her-
self with a history of atypical uterine hæmorrhages and offensive
vaginal discharge for the past two years.   A sister had died at
the age of thirty-six years from cancer of the womb.   On examination a well developed carcinoma of the cervix, extending
on to the vaginal roof on the right side, was discovered.

The curious feature of the case, and one which he desired to
emphasize, was that the patient had become very stout and
looked healthier than ever before in her life during the past two
years, in which her cancer had developed.   Perhaps a fondness
for alcohol, acquired during that time, may aid in explaining
the anomaly.

The uterus and part of the vagina were removed by vaginal hysterectomy on October 17, 1893, the operation proving unusually difficult on account of the stoutness of the patient and the impossibility of drawing the uterus down. This operation ripened in him a resolution in future to remove all similar uteri from above, after circumscribing the malignant disease from below by the proper vaginal incision.

He felt sure that he could have completed the operation more quickly and easily in that way in the present case. The patient made an uneventful recovery and was discharged two days ago. The specimen showed an advanced epithelioma of the cervix and a myoma, 2.5 centimetres in diameter, of the fundus.

CASE XI.—*Carcinoma of the Cervix Uteri. Cælio-colpo-hysterectomy. Recovery.*

The patient, a married woman aged forty-four years, was referred to him by Dr. H. Ruhl. Her left breast was removed for carcinoma at the Cancer Hospital four months before coming under the observation of Dr. Edebohls, October 22, 1893. The breast wound was well healed, with the exception of a small superficial excoriation at one point, and there was no evidence of a return of the disease.

On examination of the pelvic organs the uterus was found of proper size, but retroverted and fastened by adhesions to the rectum.

Three or four distinct malignant nodules could be palpated in the cervix. On October 23, 1893, the uterus was removed in one piece with the tubes and ovaries by combined abdominal and vaginal hysterectomy, the operation, in spite of adhesions, consuming but little more than half the time and proving much more easy than in the case just detailed, in which the uterus was removed *per vaginam*. The patient is to-day, fifteen days after operation, ready to leave bed.

Diagnosis of carcinoma of cervix confirmed, after examination, by Dr. Freeborn.

CASE XII.—*Suppurating Intraligamentous Cystoma. Abscess of Left Ovary. Double Pyosalpinx. Ovariotomy and Cæliohysterectomy. Rupture and Suture of Large Intestine. Death.*

M. M., a married woman aged thirty-nine years, was seen in consultation with Dr. George F. Carey. She had been ill with very distressing pelvic symptoms for three years past, and one year previously had been advised to have an operation for a tumor mass in the pelvis. On examination a large tumor mass, in which uterus, ovaries, and tubes were imbedded beyond recognition, was found filling the pelvis and lower part of abdomen. The tumor mass was of such density in parts that the diagnosis of fibromata complicated by inflammation of the appendages and of the pelvic peritonæum was hazarded. The patient was exhausted in the extreme by prolonged, intense suffering, by inability to retain food upon the stomach, and by loss of sleep. Her days were numbered unless operation could bring relief.

Operation, October 19, 1893. After curetting the uterus and disinfecting both it and the vagina, the abdomen was opened above the pubis, when the following conditions, were one after another revealed : Extensive and most firmly organized adhesions of the intestines to each other, to all the pelvic viscera, and to the omentum ; a moderate-sized abscess of the left ovary ; two good-sized tubes full of pus ; and an intraligamentous cystoma, ten centimetres in diameter, the contents of which had undergone purulent changes. The cyst was enucleated and removed entire without rupture, as was one of the pus tubes. The second pus tube and the ovarian abscess were, however, torn into and the peritonæum defiled with pus, which was at once removed by irrigation.

The uterus was removed mainly to secure free drainage downward into the vagina, the necessity of which became apparent soon after opening the abdomen. It was removed prior to removal of the appendages and of the cyst, because the latter structures could be reached better after the uterus was out of the way, and because, in washing out the pelvic cavity, it was considered desirable to have a free outflow for detritus and fluid downward through the vagina. In separating adhesions, an irregular rent had been torn into the large intestine. This was closed by a double row of sutures—one for the mucosa and a second continuous Lembert suture for the peritonæum. Packing the pelvis with gauze, the end of which led down into the vagina, drawing two rubber drainage-tubes through from the lower angle of the abdominal wound to the vagina, and closing the abdomen, completed the operation, which lasted two hours. The enfeebled patient never rallied from the state of shock, in which she died eleven hours after operation.

CASE XIII. *Cœlio-hysterectomy for Chronic Metritis. Recovery.*

F. M., aged forty-two years, married, was referred to him by her family physician, Dr. Alexander Strong, in July of this year. On July 17, 1893, Dr. Edebohls operated for the radical cure of an irreducible umbilical hernia, the hernial contents being composed of omentum and small intestine. Patient bore the operation well, made a good recovery, and remains cured of her hernia to this day. She suffered, however, from an aggravated chronic metritis, with extensive laceration of the cervix. The uterus was large, flabby, retroverted, and measured ten and a half centimeters in depth by the sound.

On August 8th, the cervix was removed by amputation close to the vaginal vault, and the uterus was thoroughly curetted. When the operation was finished the length of the uterus was found to have decreased about four centimeters, the sound now indicating but six and a half centimeters of depth. It was hoped by these procedures—the curettage and the amputation—to favorably influence the chronic metritis and to start a healthy involution of the uterus.

Instead of this, atypical uterine hæmorrhages made their appearance, and on October 20th the uterus was found to have again greatly enlarged, its cavity now measuring over eleven centimeters in depth. Under these conditions total extirpation of the uterus was advised and accepted by the patient and her physicians.

On October 24th, 1893, the uterus was removed *via* the abdomen. The omentum, small and large intestines, were found adherent to all the surfaces of the uterus, to the annexa, and to both faces of the broad ligaments. The adhesions were very dense, requiring great care in their separation, a full hour being consumed in this work before the uterus was sufficiently free to proceed with its extirpation. Some specially dense adhesions of the left ovary could not be separated, and a portion of this ovary had to be left attached to the bowel. The removal of the uterus proved difficult, owing to firm fixation of the cervix, which latter had to be removed in sections. The peritonæum was closed across the floor of the pelvis, and the abdomen closed without irrigation and without drainage.

The universal adhesions of the intestines and the fixation of the cervix rendered the operation extremely difficult and tedious, the time required being two and a half hours, the longest time he had ever spent over any operation. The patient suffered from fatty degeneration of the heart, which was a source of great anxiety for a week following operation. To-day, two weeks after operation, she is safely convalescent. (Complete, though slow recovery.)

The progress of the case under observation, as related above, gave the indication for the operation of total extirpation of the uterus.

CASE XIV. *Cœlio-hysterectomy for Fibroma. Recovery.*

Neither the history of this patient, a married woman aged forty years, nor the specimen—one of multiple fibromata of the uterus, one of them intraligamentous—presented anything of special interest.

The uterus was removed on October 31, 1893, by cœlio-hysterectomy performed after the method described by Dr. Edebohls in a paper read before the recent Pan-American Medical Congress, (*American Journal of Obstetrics,* November, 1893), as, in fact, were all the other uteri removed by abdominal hysterectomy, with one exception, which he had presented this evening. This one exception (Case xii.) did not admit of typical extirpation of the uterus after his method.

He desired to draw attention to one other point before closing, which was that he had about reached the conclusion that in all cases, except those of prolapsus, in which extirpation of the uterus was called for, the operation was best done from above. If performed for malignant disease of the cervix, he would first circumscribe the disease by an appropriate vaginal incision, then open the abdomen and remove the uterus from above. If for non-malignant disease, *all* the work should be done from above.

(The statement made above, that the specimen presented nothing of special interest, must be modified, inasmuch as Dr. G. C. Freeborn, upon examination of the uterus, found in addition to a number of fibromata, an angio-sarcoma with extensive hyaline degeneration of the arteries of the malignant neoplasm.

This, in Dr. Freeborn's opinion, renders the specimen an exceedingly rare, if not unique one. The results of the microscopical examination will probably be published, more *in extenso,* elsewhere.

The intraligamentous tumor, taken at the time of operation for a fibroma, proved to be an outgrowth of the angio-sarcoma into the broad ligament. Dr. Freeborn reported that an unbroken capsule surrounded this neoplasm, and he would therefore say that it had been *entirely* removed.

An examination of the patient, ten weeks after operation, fails to disclose any evidence of recurrence.)

REMARKS UPON REMOVAL OF THE UTERUS IN DISEASES OF THE APPENDAGES.

(Made in the discussion of a paper by Dr. Polk.)

[From the Transactions of the New York Obstetrical Society, November 7, 1893.]

DR. GEORGE M. EDEBOHLS found himself unable to assent, in its entirety, to the broad proposition of Dr. Polk to remove the entire uterus in every case of disease of the uterine appendages in which such diseases called for removal of both tubes and both ovaries. His objection was not founded upon the fact that he considered total extirpation of the uterus in such cases a difficult procedure, or one adding much to the gravity and danger of the operation ; nor was it based upon theoretical considerations of the value of the uterus as an essential factor in the preservation of the arch supporting the pelvic floor. He had undergone much the same evolution as a gynæcological surgeon with his friends Dr. Polk and Dr. Krug, and as a result of such evolution, and of practical experience, he had, with these gentlemen, come to regard total extirpation of the uterus by way of the abdomen— cœlio-hysterectomy—as about as safe an operation, and one as well borne by the patient, as ovariotomy, the complications, so far as adhesions and other difficulties are concerned, being the same, case for case. He wished to reiterate, therefore, that it was not the difficulty of the operation that deterred him from applying it universally in the class of cases under consideration ; on the contrary, the operation was so easy, and withal so congenial to his own surgical proclivities, that it was difficult for him to resist the temptation to apply it more frequently than he did.

He had had opportunity twice to make post-mortem study of the bodies of women in whom the entire uterus had been removed some time previously, in one by vaginal and in the other by abdominal hysterectomy. He had in both instances subjected the pelvic floor to critical examination and tests as regards its strength, and felt fully satisfied that the absence of the uterus in no way impaired the latter. This was only corroborative of our daily experience on the living woman. He had as yet failed to find, among the now fairly large number of patients whose uteri he had for various reasons removed, any indications of a weakness of the pelvic floor, nor did he recollect any allusion to an *actual* occurrence of this kind in the literature of the subject.

At a meeting of this Society, held February 16, 1892, Dr. Polk presented specimens from a case of pyosalpinx and suppurating ovary, in which, after removing the diseased appendages, he found the uterus had received so much injury and was bleeding so freely that he removed the major part of it, *leaving a stump of cervix in situ*. While listening to Dr. Polk's words on that occasion several cases of his own, in which he had performed cœliotomy for suppurative inflammations of the pelvic organs, came to mind, some of which he thought would have been better dealt with by total extirpation of the uterus instead of simple removal of the diseased appendages. Two weeks later—on March 1, 1892—Dr. Edebohls presented to the Society the specimens from a case of puerperal pyosalpinx and intra-peritoneal abscess, with chronic pelvi-peritonitis, which he had successfully treated by removing from above the *entire* uterus, together with the diseased tubes and ovaries, in one piece. The indication in his case was the inability to distinguish, with the abdomen open, where the uterus ended and the annexa began. While Dr. Polk's case suggested the idea, his own was, as far as he knew, the first reported case, in this country at least, in which the *entire* uterus had been removed *per abdomen*, together with the diseased annexa, for suppurative disease of the pelvic organs, the uterus itself not being materially diseased.

He had since then operated in the same way upon three additional cases, specimens from the last of which he had presented in the earlier part of the evening. These four cases had constituted the most serious and desperate of the cases of intrapelvic suppuration he had encountered during the past twenty months.

While thus demonstrating, by his practice, his acceptance, within proper limitations of indication, of the principle involved in the discussion, he radically dissented from the sweeping proposition to remove the uterus in all cases in which the condition of the tubes and ovaries called for the removal of both of the latter. Dr. Polk's argument, if he understood it correctly, was that the operation of removal of the entire uterus was an easy one ; that it added nothing to the risks incurred by the patient ; that the uterus, after removal of the tubes and ovaries, was an entirely useless organ, still liable, however, to disease.   Dr. Edebohls was willing to allow the correctness of all these claims ; more than that, he was convinced of their correctness. The same line of reasoning, however, applied, for instance, to the appendix vermiformis, would call for removal of the latter whenever and for whatever purpose the abdomen is opened ; the appendix is, so far as we know, a useless organ, often dangerous to life ; its removal is easy and does not add materially to the dangers of the operation, whatever it be, for which the abdomen has been opened.

Just as soon as we reached universal acceptance of the principle that the *healthy* appendix vermiformis must be removed whenever we open the abdomen for any purpose, just so soon would he be ready to extend the principle to the uterus, and to accept the proposition of Dr. Polk to remove it, diseased or healthy, whenever we must sacrifice the appendages. When that time arrived, to be sure, a cœliotomy for the removal of the uterine appendages would imply a rather large contract ; hysterectomy and ecphyadectomy must be added to the bilateral salpingo-oophorectomy. Who dare say, however, that such a time may not come, and that Péan, Ségond, Polk and Krug are not merely a number of years ahead of their day and generation ?

For the present, when performing bilateral salpingo-oophorectomy, he was content to leave the uterus, if that organ were found healthy, or only so slightly diseased that curettage would suffice to restore its health. He *invariably* did curettage of the uterus immediately preceding every cœliotomy for diseased appendages, and, where the uterus was left after removal of the tubes and ovaries, *generally* attached the fundus by suture to the abdominal walls when closing the latter. Uteri thus treated very, very rarely gave their bearers subsequent trouble. The first indication, then, for him to remove the uterus, when performing bilateral salpingo-oophorectomy, was furnished by a diseased condition of the uterus itself, such diseased condition not admitting of removal by curettage of the uterus.

A second indication was illustrated by the specimen he had presented to the Society twenty months ago : inability to distinguish the boundary line between uterus and annexa, the genitalia interna forming a conglomerate mass impossible of resolution into its component elements ; or, in attempting such resolution, the uterus may receive so much injury, and the bleeding be so profuse, that it may be better, as in Dr. Polk's case already alluded to, to remove the uterus entirely.

A third indication may be found in the necessity of absolutely first-class drainage in very foul pus cases, such as the one specimens from which he had presented that evening. There was no doubt in his mind that the most perfect drainage possible in such cases was obtained by removing the uterus and draining the pelvic cavity through the vagina. It was mainly, if not altogether, for the purpose of availing themselves of this nearly ideal drainage downward through the vagina that Péan, Ségond, and their followers advocated and practiced vaginal extirpation of the uterus by *morcellement* in cases of the different varieties of inflammation of the female pelvic organs.

Time did not on this occasion permit of his entering more fully upon a comparison of the merits of the two procedures—cœlio-hysterectomy and colpo-hysterectomy by *morcellement*—in those cases of diseased annexa calling for removal of the uterus. That subject would on some future occasion, perhaps not distant, again come before the Society, and he would then avail himself of the privilege of discussing it. For the present he would merely state that when, under the conditions named, the uterus required removal, he would invariably prefer to remove it from above.

198 SECOND AVENUE, NEW YORK, N. Y.